Organisation Change

Japan

Second Edition

In an acknowledgement of the
outstanding work of
Edwards Deming
(1900 – 1993)

David Bannister

Organisation Change – Japan

Introduction

This Concise Guide to Organisational Change – Japan provides a quick way for you to gain an understanding of this complex subject. It looks to summarise the essence of the subject and engage you as the reader in the main areas of knowledge supporting this subject. Most importantly the objective is to encourage you to go deeper into the subject so the background on the experts (gurus) behind this subject and their books should prove invaluable. As the author my understanding was acquired by practical industrial experience, training courses, seminars and reading. Much of my understanding came from reading many books so I have a view that you should read these books for yourself to acquire the knowledge from these experts first hand in their words not mine. The Bibliography included is therefore important since it details the books valued by the author although there are no doubt many more just as effective. With, sadly, most of these books now heading for the skip they can be easily acquired at low cost over the internet from many Second Hand book providers.

Preface

The author having worked for an American manufacturing business that decided to adopt Japanese manufacturing principles, based upon the Toyota Production System (TPS), gained a unique insight into the different American and Japanese approaches to manufacturing. In fact the business also applied the American Six Sigma principles at the same time as applying TPS. It soon became obvious that success depended upon how effectively the Japanese manufacturing culture could be adopted by the workforce.

But the most unusual aspect about Japanese manufacturing culture was it was taught to the Japanese by an American called Edwards Deming. This was at a time when Deming could not persuade American manufacturing to adopt his ideas.

The idea for this Paperback and eBook evolved out of an objective to try and communicate concisely how the Japanese approach Organisation Change which is very different to the American approach. But it needs to be appreciated that the Japanese culture has its own limitations against the American culture that more readily accepts radical changes. It is Japanese evolutionary change verse American revolutionary change. In fact those that choose to combine the two approaches are going to be the most successful businesses of the future.

So what is this Organisation Change?

So if we are going to make a change what was there before the change? Manufacturing had prior to this change been based upon two very similar scientific ideologies. One European called Taylorist and one American called Fordist. Essentially they both defined the principles of mass production. Taylorist being a notion of modularity and the division of labour. Fordist being the use of standard parts that fitted together exactly and simple standard human work steps that made or assembled these parts.

So mass production achieved by design the simplicity and interchangeability of both the parts and the labour steps to make these parts. The essence was to make and assemble strictly against the clock with the human labour used only to focus upon achieving the production rates. In terms of quality this was achieved by a separate set of inspectors accepting or rejecting the products produced. Good parts were kept and bad parts were repaired or scrapped.

The workers where in fact unskilled trained to only operate in what would now be termed a robotic way. They were denied involvement in the work they did or the finished products they created. The workers ingenuity, creativity, enthusiasm, imagination and commitment were supressed by the bigger standardised industrial system.

So the organisation change described in this guide takes manufacturing from mass production to lean production. In fact before mass production manufactured goods were hand crafted in small batches using the specialist skills of the workers. One view is that lean manufacturing has introduced back into the workplace some of the best aspects of a craft industry approach particularly in respect of the worker taking ownership of the work being undertaken. But the principles of interchangeability of parts and standard ways of working common to mass production were retained and often enhanced. But the workers were integral to the design and achievement of these mass production principles and not just made to undertake them.

Forward

Japan has become one of the most effective manufacturing countries in the world and its approach to manufacturing has been copied throughout the world. But it was not always the case. In fact prior to the Second World War Japan was not a particularly effective manufacturing country. In fact it had a reputation for a lack of innovation and was always accused of only being able to copy other people's designs.

It was in the post Second World War era when things changed with much of this change credited to an American called Edwards Deming who was seconded to assist in the rebuilding of Japanese industry. Deming had formulated many new concepts as to how to improve manufacturing. But in America he failed to get his radical new ideas recognised and implemented. It was the Japanese that were the first to apply these concepts to manufacturing which then over a 5 year period catapulted Japan to be become a leading manufacturing nation.

You should never proceed with any Change Management initiative without understanding some of the background to Edwards Deming's impact on the Japanese and later the American manufacturing revolution. His teaching were holistic or 360, and some thought peculiar, long before these terms ever existed. This is because his teachings ranged from statistics to psychology to physics to engineering.

Edwards Deming was born in 1900 and he obtained his first degree in 1921 staying on after to teach mathematics and engineering later going on to teach physics. He developed a keen interest in mathematical statistics. But it was the work of Walter Shewhart, the originator of the concepts of statistical control over processes and the concept of control charts in industrial production processes that fired his imagination and enthusiasm. The concept of controlled variability dependent on the original process design and how people are trained on undertaking the processes verse uncontrolled variability from sources outside the process was a paradigm shift in thinking. The essence was process design was a critical factor in controlling quality. The view developed that people should work the system that management should have originally carefully designed and now should be continuously improving. The concept evolved that part of the management team should always be exclusively dedicated to this process design and improvement aspect.

The other guru that worked alongside Edwards Deming was Joseph Juran who established that 85% of problems (variation) was due to management and only 15% due to the workers. Deming revised this to 94% and 6% respectively. It was Deming that extended what had predominantly been manufacturing based thinking across to logistics, service operations and office work.

Deming had a really low profile in America. After the Second World War he was invited to Japan by General MacArther to be an advisor to the Japanese Census. Whilst there he meet members of the Japanese Scientists and Engineers (JUSE) which had been formed in 1946 to support the reconstruction of Japan after the devastation of war. A delegation from Bell Telephone Laboratories showed the Japanese some of Deming's work on Quality Control. He was asked to lecture to them on statistical methods for industry in 1950. The Japanese listened astonished as Deming told them that if they adopted his methods then they would begin to capture world markets in a very few years.

Deming in his teaching went beyond the purely mathematical concepts of Total Quality Control (TQC) with teachings like the "consumer is the most important part of the production line" and that they should engage with their suppliers as partners in the supply of quality components bringing them closely into the organisations production boundaries. This was more about the psychology of human behaviour than mathematics. In fact he taught the techniques that could be employed to improve the statistical performance.

The Japanese culture particularly their strong enthusiasm to learn and their dedication to a detailed rule based approach allowed them to fully embrace Deming's ideas. Within 5 years they had become one of the most effective manufacturing countries in the world and started to trigger the decline in Western manufacturing economies.

It was 1974 before America started to wake up to the teachings of Edwards Deming by which time Japan had evolved the Deming teachings into a powerful set of methodologies with Toyota the most significant leader in the use of the Deming methodology evolving it into its own Toyota Productions System (TPS).

It was the President of the Nashua Corporation, who whilst in Japan, noted that Ricoh was undertaking a five year programme to obtain the best known industrial award in Japan. That award was the Deming Prize. The Deming Prize awards were even broadcast on prime time television. It was 1979 before when Deming who was 78 years old that he was asked to help this first American business Nashua with his teachings.

Deming had worked with Japanese industry for 30 years before American industry started to take note of his new approach to manufacturing. What could be summarised as a statistical control approach to manufacturing rather than just mass production. Japanese success had been based upon statistical quality control introduced to them by an American. It was the NBC television documentary "If Japan Can, Why Can't We?" when broadcast in June 1980 that in fact launched Edwards Deming onto the American manufacturing scene. Suddenly Deming was in great demand in America with his celebrated four day management seminars.

The strange thing being that Deming was presented with the American National Medal of Technology in 1987 by President Reagan 27 years after in 1960 he had been decorated in the name of the Japanese Emperor with the Sacred Treasure being the first American to receive such an honour for his contribution to the Japanese manufacturing industry.

It is important to understand Deming's thinking and although his 14 Point Strategy may appear a little dated in our new digital world it is worth reading these points in their original wording since these points have become the foundation of all modern Organisation Change thinking. In reading them you will soon appreciate the focus on people and the motivation of people.

Deming's Out of Crisis 14 Point Strategy

01 Create constancy of purpose for improvement of product and service
02 Adopt the new philosophy
03 Cease dependence on inspection to achieve quality
04 End the practice of awarding business on the basis of price alone. Instead minimise total cost by working with a single supplier
05 Improve constantly and forever every process for planning, production and service.
06 Institute training on the job
07 Adopt and institute leadership
08 Drive out fear
09 Break down barriers between staff areas
10 Element slogans, exhortations and targets for the workforce
11 Eliminate numerical quotas for the work force and numerical goals for management
12 Remove barriers that rob people of pride of workmanship. Eliminate the annual rating or merit system
13 Institute a vigorous program of education and self-improvement for everyone
14 Put everybody in the company to work to accomplish transformation

Elements of Organisation Change

Introduction

Since the 1980's organisations of all kinds have undertaken organisational change under a variety of names including restructuring, right-sizing, de-layering, out-sourcing, down-sizing, business transformation and process re-engineering. In the 1990's we have moved onto Lean Management, Six Sigma and Toyota Production Systems. (TPS) These are some of the names used to describe the process of Change Management and generally most workforces are not impressed when management announce the commencement of such a change process. This is usually because they fear for their jobs and usually this is a justified fear. With the digital revolution often leading this need for change the effect on being able to reduce the number of jobs in an organisation is often dramatic.

Change is essentially about for those that remain within the organisation a process of unlearning and relearning. In an organisation it is often dependent on a co-ordinated change process across many employees and different functional areas. Change is as much about process changes as behavioural changes. It is the behavioural changes that are the most difficult.

Anyone installing the same new computer application software into different units in an organisation will tell you about the unit that thinks it is great and the unit that thinks it is a disaster. The units do the same job and it is the same software but the difference in the acceptance is down to the behaviour of the staff in the unit. A good manager selling the software benefits in the unit to other staff will make all the difference to the final acceptance of the software by the unit.

The main behavioural element is those who initiate the change believe it is worth their efforts. The change has to create enthusiasm in those looking to implement the change.

Currently (2019) change increasingly in organisations is technology lead and this pace of change is accelerating. Organisations have to increasingly be flexible to sudden changes that need to be made to ensure the survival of the organisation. To survive organisations need to continually adapt and change to the dynamics of the environment within which they operate.

One of the most radical and effective changes an organisation can do is to install a completely new computer application system. After a correctly managed project inclusive of functional analysis at the start and comprehensive user training at the end this can allow for an organisation to change its working practices and processes overnight. This approach can be termed the revolution style of change rather than an evolutionary approach. This revolutionary approach brings about a sudden change in an organisation's business processes which is difficult to achieve any other way. But the revolutionary approach can result in all the issues and problems surfacing after the go live date.

This brings us to the Japanese approach termed kaizen which means continuous improvement. Kaizen is a totally holistic approach across a whole organisation and out to clients, customers and partners. It views every contributor as part of a team with the customer or client inclusive in this team. Within the organisation the workflow is analysed across functional boundaries looking to continually remove functional interface problems whilst always focussing on the needs of the customer. The approach is always inclusive of waste reduction including the reduction in work steps and work step documentation creation.

One of the constant areas of discussion in terms of how to make change is whether it should be evolutionary or revolutionary. The American approach tends to be revolutionary whilst the Japanese approach is evolutionary. Although this is a too simplistic statement. Effective organisations mix and match the use of both of these approaches to change having an awareness that both types are needed but that the evolution approach must always be ongoing. Revolutionary change is factored in to meet specific objectives on top of this continuous evolutionary change process.

But the Japanese take it deeper than western organisations in that they consider kaizen to be a change in the organisations belief systems. The Japanese refer to the need for people to believe things other than what they now believe. Western organisations call this having a vision and importantly then communicating this to make it a shared vision. Behaviour then has to change to support the new beliefs or vision. But whereas western organisations treat things as projects in that they get completed the Japanese have as part of their belief system the need to be continuously looking for a better way of doing everything.

The significant point is the beliefs have to be changed first before changes can be made. The word belief really does define the passion and enthusiasm needed to make changes. Those leading the change have to really believe in the change they are making showing complete dedication. The Japanese look for all their leaders to have excellent leadership qualities with a great belief in leading by example. Leaders are encouraged to get directly involved in the frontline work of the organisation and in particular in product usage and assessment. Stories of Japanese car plant executives walking down regularly to drive and test cars directly off the production line are part of Japanese manufacturing folklore.

Managing change in western organisations tends to be project orientated often commenced with an exercise to change the culture and commitment of the organisation to one that will actively support the change process. This is the first error in that changing culture and commitment takes years not months. Change has to have a sponsor being a Senior Change Leader that can commit resources to the change project and ultimately be responsible for the success or failure of the change initiative. Importantly the Senior Leader will appoint a Change Leadership Team. The Senior Change Leader and the appointed Change Leadership Team will need to communicate the change vision. The vision defines a clear picture of how the future will look after the change process has been completed with the clear communication of this vision critical to the success of the change process. The change process then needs planning with tasks, budgets and accountabilities defined. Progress needs monitoring against the plans in terms of tasks completed and expenditure against set budgets. In large organisations change maybe completed in one location or unit first so it can be refined before a roll out across the whole organisation. The roll out maybe completed using eLearning or Trainer lead knowledge transfer techniques.

Let us consider the main elements of Change Management.

1. Organisation
2. People
3. Culture
4. Business Processes
5. Metrics
6. Standards

1. Organisation

Organisations use people to operate business processes with these people organised into functional areas that are managed by a hierarchical management structure.

Purpose

Sometimes called a mission statement it a concise description of the purpose of the organisation. Under Japanese thinking this is focussed upon the belief systems of the organisation.

Vision

The vision defines what the organisation wants to become in the future. This needs to both quantitative measurable elements and qualitative aspirations. The vision can be inclusive of products and services that may be developed in the future. This maybe be complimentary to what currently exists or the vision maybe inclusive of totally new product and service offerings.

Values

This is defined as the code of conduct that people need to operate within the organisation.

Products or Services

To understand what the organisation delivers and to understand how customers or clients perceive the product or service they are receiving.

Differentiation

The differences between ourselves and the competitors. In which market segment do we operate? Top Quality at Top Price, Medium Quality at Medium Price or Low Quality at Low Price. Although admitting to Medium Quality or Low Quality would be a major marketing mistake whereas just focussing on Medium Price and Low Price would be sufficient.

Structure

Organisations historically tend to divide work up into functional areas. So Design Departments, Sales Departments, Production Departments, Logistics Departments represent different functional areas. Structure has a huge impact on how effective an organisation operates. The organisation structure defines how an organisation divides up the work. If work is viewed as a workflow then how is this workflow channelled through the organisation domains be they called Divisions, Departments or Functional Areas. How does each of these domains add value to the workflow? How is the vertical workflow within these domains organised in terms of management hierarchy?

The more complex the structure particularly along the path of the workflow the slower the organisation will operate. More complexity equates to more costs. In terms of meeting customer or client needs the more complexity makes it less responsive. Having domains effects organisation politics with this wasting time and effort so not adding any value.

One trend is to even for very large organisations have smaller tightly controlled operating units with an increasing trend to outsource non-core activities to third party operators. This trend creates increasingly complex structures often termed networked organisations. Networked Organisations are more focussed on meeting specific maybe local objectives but by their nature they are more difficult to centrally control.

Hierarchy

Hierarchy defines the ranks of seniority in the organisation. The purpose of a hierarchy is to define where and at what level decisions are made within the organisation. It establishes the responsibility and accountability of roles within the organisation. The current change management trend is that large hierarchies are slow, costly and ineffective. Flatter organisations with fewer layers of hierarchy is a current important change management objective.

An organisation strategy is the process of setting out a vision for the organisation so it succeeds either competitively against other organisations or it meets its defined objectives if not in the business sector. Establishing an organisational strategy requires going through a number of steps to define a number of specific elements. One of the most major changes that can be made to an organisation is to change its management structure. A management structure defines the decision making capabilities of the management roles within an organisation.

The management structure defines who supervises who within the organisation. It defines the social framework of the organisation.

Traditional organisations have characteristics that all employees have come to see as the normal organisational structure. These characteristics include top down goal setting, each employee having a single superior, vertical communications, power defined by remuneration grading levels and finite manager boundaries of control. This lends itself to the classic corporate culture and politics particularly as employees try to get promoted through the hierarchy. Changes from this traditional model have been led by the very creative and demanding technological and media industries where employee motivation and retention is critical to their success. It is not surprising that it is the technological industries that have used technology to remove management layers. Information flows "first hand" from the bottom to the top of the organisation via email or collaborative systems. The removal of layers has resulted in decision making moving down the organisation hierarchy nearer the sharp end of the organisation.

Now this devolution of decision making has to be a controlled process and importantly you have to define what decisions are allowed by what role within the organisation. So decisions have to be quantified in terms of scope or value or impact. This requires the need for a systemic approach to decision making.

Taking the sharp end of the business as being Level 1 then at which level does the Chief Executive sit in the hierarchy? In the West it was commonly at Level 4 but the trend is to bring it down to Level 3. In government organisations it could be at Level 6 or higher.

Within the hierarchy the Level 2 or Team Leader level is where most of the management decision making is devolved. One level up from the sharp end where the work is actually done this is a crucial role.

Business Processes

Business processes define the way work flows through an organisation. Each business process should add value to the work as it moves through the organisation. Each process whilst adding value needs also to ensure quality is maintained. Business processes can apply to the manufacture of a product or the delivery of services or administrative activities.

2. People

Organisations are operated by people. Organisations differ in the degree to which the workflows are automated and so less dependent on people. People within an organisation are at work trained to operate within the set culture of the organisation. The most difficult aspect of change is moving people through the change process. Human beings like stability and continuity in fact they are most comfortable when no change is taking place. Change means the effort of re-learning, the risk of your job changing or disappearing. Change poses a threat and the future becomes unknown. But it has also to be acknowledged that some people thrive on change finding it a key stimulant in their working lives.
Edwards Deming when asked what he had learned after many years of teaching about change and quality management replied simply "people are important." Change thinking has come to acknowledge that the psychodynamics of individual change parallels closely what happens in an organisation which is undergoing change. The organisation has the same characteristics as an individual as it goes through a change process. This is not surprising since because organisations are social systems they are in fact made up of people. One aspect of organisational change is

the higher in the organisation the change leader is placed the more effective that person will be in changing the organisation.
Many models exist that represent the so called cycle of change by they usually follow the format below.

Stage 1 Negative reactions to proposed change.

Change resistance is an uncomfortable period in the cycle. It includes negative feelings and anxiety. The key is to get people believing that the consequences of not changing will ensure more problems in the long term than those that will be experienced by making the changes now.

Stage 2 The defining event.

The "wake up call" or trigger to commence the change process. The emotional energy that went into resisting the change is now re-directed into thinking of making the change. Sometimes called the "being between trapezes" phase where you are in mid-air having left one state and about to enter another.

Stage 3 Making the commitment.

Changing over and voicing of the commitment to change. Broadcasting the intent to leave the status quo. It can be a motivation phase with people reinforcing other people of the value of the change. Spirits rise and wellbeing returns.

Stage 4 Self Discovery

Those having experienced the above cycle gain a better self-knowledge and self-awareness. Reflective thinking allows self-evaluation of the process as a learning experience.

Stage 5 The Norm

The new changed environment now becomes the norm. The past is now forgotten and the new environment becomes the new status quo.

3. Culture

In organisational terms the culture defines the beliefs held by employees of the organisation as to what are the prevailing values within the organisation. Culture is not established by dictates it is a product of observed behaviours particularly those communicated by the management and knowledge transfer roles. Change itself is a cultural characteristic. Organisations can have within their culture an enthusiasm for change and a frustration when change is absent. Changing an organisations culture is difficult because it is established over a long period of observed behaviour. The short cultural change training course achieves nothing. Cultural change takes years to implement. Culture as a set of beliefs shows itself in the attitudes and behaviour patterns of the people within the organisation.

Organisations can make claims about their culture often at a management level where in practice these are not actively practiced at the front line operational level. When there is this disconnect between what management claim and the actual practice then staff can become cynical about leaders who say one thing and do another. The real culture and values of an organisation are always there to be witnessed as can be the differences between those being practiced and the management's perception of the stated values they believe are being operated. In assessing an organisation's culture just ask a cross section of staff from top to bottom to describe the culture of the organisation. Avoid the standardised notice board or newsletter projections of culture and establish what it is really like in practice.

Culture has a much broader connotation in that races, countries and continents can be viewed as having different cultures. You can even end up with an island race culture resulting in comparisons between countries like the United Kingdom, Japan and New Zealand. You can also have Western and Eastern cultures. Culture is considered to be a group specific behaviour that is acquired, at least in part, from social influences. Culture can also be defined by generations normally defined in 10 year age groupings. In particular the acceptance and usage of technologies is very generation based often linked to the skills and aptitudes acquired in the formative years.

In organisational terms the Japanese culture naturally focuses upon the success of the group rather than that of the individual. The Japanese are natural team workers with a tendency to accept and adhere to authority rather than challenge authority. They patiently listen to inputs from all group members showing considerable respect for each other. At the other extreme the American culture is based upon the success of the individual. In school, college or university settings the students are encouraged to challenge any teachings. Americans are taught from an early age to challenge and get their point across forcibly and in an argumentative style . This approach strengths individual thinking and is deemed to support thinking outside of the box and therefore innovation. Unfortunately as a style it can generate both major seismic innovative shifts but it can generate enormous failures.

Effective organisations need a delicate balance between these different cultures. The need to acknowledge that major innovation often comes from an individual whilst success in the long term is based upon the team working together constructively on the implementation of the innovation. Some organisations have deliberately setup the creative functions separately from the day to day operation functions. The creative function has to create and the operation function has not to create but just operate what has been created. This clear separation can create a particularly effective organisational model whilst capitalising on the best of both of these cultures. In the application of cultural change to an organisation each culture, creative and operational, are developed through different change processes. But importantly they are linked together in a tight feedback loop whereby operations can feedback to or use the creative resources. Creativity can cover product, services, processes and procedures. Operations will work on or apply all of these and feedback on all of them. Importantly there has to be a documented communication process between the creative functions and the operational functions. This these days will be a computer based solution providing this interface with collaborative working encouraged.

One of the key teachings by those looking to implement cultural change is it is not about management but it requires leadership. It also requires leadership commitment and a urgency to build up the momentum of change. Many believe that when examining culture one of the key elements for an effective culture is commitment. Commitment in an organisation is a measure of the motivation of those in the organisation to make change happen and deal with all the positive and negative behavioural issues that change generates. Many organisations lack commitment when the negatives evolve and back off from making the change. Commitment is about dealing with the negatives often by increasing the involvement and influence of those pursuing negativity. Some of these negatives may have real relevance to the change process and the process may need adjusting to take account of these issues. Staff who originally had a negative stance can become the most committed to the change process when they appreciate they have an input into the process.

Cultural change depends increasingly upon those having commitment to the change converting others in what is termed today as a "networked" or "viral" style of what amounts to cultural marketing. You have to market cultural change. The old change models restricted involvement in change to a few staff on the basis that this made it happen more quickly. But this failed because to gain commitment you need staff to have involvement and a cultural change will not work without commitment. Involvement gives those involved the feeling that they are influencing the cultural change process and the more they influence the higher their commitment.

American companies focus upon trying to implement cultural change through things like VALUE teachings. "V" for being Very efficient, "A" for A customer focus, "L" for Lots of teamwork, "U" for understanding and "E" for Excellence. This approach shows the sometimes thin veneer of cultural change in western businesses whereas it is totally embedded in a Japanese business.

4. Business Processes

Processes have an input, something is then done and then it produces an output. This output then becomes an input to another process. Stringed together either sequentially or in parallel or along different paths in total they make up a workflow through the organisation. The workflow can flow across different organisation functional boundaries. All organisations be they manufacturing, service or government have workflows running through them.

Workflows are lateral as they run across an organisation. These can trigger some vertical workflows moving up the organisation's hierarchy to acquire some decision making. Hoskin kanri is a vertical process where a lateral workflow has triggered a vertical need for decision making.

Process models depend upon people and technologies. Included in the term technology are the computer devices and the software that runs on these devices. The trend is increasingly to make the use of these computer devices transparent to the user with the terms pervasive or ubiquitous being used to describe this approach. There are two ways to achieving this ubiquitous objective. Firstly no human involvement with computer devices just working with other computer devices without people operating anything. Secondly making this use of the computer device requiring a human input so natural that it forms no barrier to the person undertaking the process. Just think of the way you now withdraw cash from an Automatic Tilling Device (ATM) or "hole in the wall". This is quite a complex process for which you are never trained. These days the way you interact with your smart phone to do any variety of tasks. The designers of these systems are trying to simplify and emulate the most natural ways users you want to interact with them.

Now anything that has many steps needed to achieve and end outcome often with intermediate steps depending on completion by others requires a carefully designed process model. The designers of these process models are always looking to achieve the start to end as quickly and efficiently as possible. Efficiency inevitably means minimising the human inputs to the process model with the computer to computer model being used whenever possible.

5. Metrics

In terms of Change Management there is one aspect that sits almost outside the scope of Change Management but is in fact integral to the process of change management and that is metrics. Metrics is the commonly used term to describe any process analysis using mathematics or put simply working with numbers. It has evolved from the area of Statistical Quality Control (SQC) where in its purest form any process inputs and outputs need to be measured to determine the reliability of the process. Reliability is about producing the same outputs consistently without variation so these depend upon the inputs to the process being consistent.

Put simply everybody in an organisation from top to bottom needs to collect data and plot this data on charts called control charts. Then based upon these charts they fix things which then improves the readings on the charts. This cycle of data collection, charting, fixing, data collection, charting and fixing just keeps repeating itself over and over again.

Organisations reporting on agreed metrics consistently outperform organisations that do not have a metrics culture. In terms of change management defining the key organisational metrics prior to the change and reviewing these post the change is a key activity undertaken as part of the change process.

When a customer buys a product or receives a service they have an expectation that it will meet their requirements. These requirements are what we mean by quality. Statistical Quality Control (SQC) is a collection of methods for measuring, monitoring, controlling and improving quality. The trend is to have the metrics or information distributed on an on- line basis so it can be continually reviewed and monitored whilst the process is in progress.

Metrics relating to process management are about measuring the degree of random variation in the inputs to a process so control can be exerted over the output. The aim of Statistical Process Control (SPC) is to monitor the variation in the quality measurements to detect when the process shifts to an out of control state. So the most powerful statistical process control tool is the control chart. A control chart shows the plot of a measurement against a time line. The chart will have a centre line representing the average or mean of the process when it is in control. When the process is in control the sample measurements will vary randomly around this centre line. On the control chart will be an upper line which is the upper control limit (UCL) and a lower line being the lower control limit (LCL). When the process is in control most of the sample measurements will lie between these upper and lower lines. A reading outside these limits suggests that the process might be out of control. Control charts have patterns to them and experienced users of these charts get familiar with the patterns.

This variance from the mean to these upper and lower control limits is called the standard deviation and this is represented by the Greek letter "sigma". If the standard deviation were larger the upper and lower control limits would be further from the centre line. This means that when the normal variation of the process is high it means the sample can deviate more from the centre line without suggesting that the process is out of control. We end up with control limits termed 2-sigma or 3-sigma. So 2-sigma means the probability of a sample mean lying outside the 2-sigma limit is 0.046 or 4.6% or about 1 in 20. So 3 – sigma the probability moves to 0.0027 or 0.27% or 1 in 370. Within the process world the term Six Sigma has become a legend as a standard to be aimed at by all organisations. A Six Sigma process moves to 0.00034 or 0.034% or 3.4 defects per million.

The use of the Six Sigma term to define a methodology to improve quality was commenced at Motorola in the 1970's. In fact as a methodology it followed the work of the pioneers such as Shewhart, Deming, Juran, Ishikawa and Taguchi. As a methodology it had a focus on decisions made upon verifiable data so it is very metrics focused. It also introduces the concept of "champions" who are called "Black Belts", "Green Belts" and so forth. Six Sigma has now evolved to be termed Lean Six Sigma to acknowledge the trends towards lean thinking.

Although the use of control charts is the most popular metric analysis method there are many other techniques. The Japanese refer to them as the Seven Tools of quality control which are trained in schools and used throughout organisations. In Japan they are sometimes called the Isikawa Seven. These Seven Tools are:-

1   Control Charts
2   Pareto Charts
3   Cause and Effect Diagrams
4   Checklists
5   Stratification
6   Histogram
7   Scatter Diagrams

1. Control Charts

Control Charts are used to monitor, control and improve performance over time by studying variations and most importantly the source causing these variations. They allow for monitoring and distinguishing so called normal variations sometimes called noise from special out of control variations. Control Charts are easily read and understood with them allowing for excellent open discussions as to what is going on and how things can be improved.

2. Pareto Charts

Pareto charts focus upon the key problems. They are used to focus efforts on the problems that offer the greatest potential for improvement by showing their relative frequency or size in a descending bar graph. The Pareto principle is that 20% of the sources cause 80% of any problem. Pareto Charts are easily understood with the way they are presented often having a high visual impact on the receiving audience. They are excellent for creating focus and fully support the quick wins approach to improving performance. Before and after Pareto Charts are an ideal way of representing the effect of changed processes.

3. Cause and Effect Diagrams

Cause and Effect Diagrams (Fishbone Diagrams) allow the identification and graphical display often in increasing detail of all the possible causes related to a problem or condition to allow for the discovery of the root cause or causes. They are an excellent and engaging way for a team to collaborate on the solution to a problem in so called brainstorming sessions. For service processes it covers the analysis of policies which are high level decision rules and procedures which are low level detailed steps in conjunction with computer systems and staff.

4. Checklists

Checklists can be used as a way of counting occurrences either historically or by a current observation process. Works best when there is a standard list of events or conditions that are being checked. Getting the checklist to include the correct parameters being monitored in the first place is critical. Car servicing checklists are examples of how a checklist can work especially when marked up with the failures. This is a good example to highlight the importance of frequency of doing the check. Should it be weekly, monthly or annually. In fact the checklist will be different for each frequency period based upon the likely failure rate of each parameter being tested.

5. Stratification

Stratification as the name suggests is about selecting from a data population in a structured or layered way. It is about basing sampling on defined groupings. Consider a political survey. If we wanted to represent the diversity of the population then sampling of each group would be in proportion to the group's overall proportion of the population. Just random sampling would fail to reflect correctly in this diversity aspect.

6. Histogram

Histograms summarise data from a process that has been collected over a period of time with this graphically presented as a frequency distribution in a bar form. Ideal for showing the frequency of occurrence. Visually effective in that they show themselves as a variety of different shapes. The normal distribution (bell shape) being the most common. Although depending on the parameter being monitored many other shapes can exist. Consider call rates where calls per hour are plotted with time horizontally and number of calls vertically. This can have many peaks and troughs over time. Histograms are very similar to Control Charts.

7. Scatter Diagrams

Scatter diagrams allow for the visualisation of the relationship between two different variables. Importantly they can be used to establish if a relationship exists. When the scattering is completely random there is no correlation. The clustering of the points along a straight line can show a positive correlation. As a technique it is often used as a follow on to Cause and Effect Diagrams to prove a link between the cause and the effect.

Obviously there are many other techniques that can be used but the Japanese are very good at focusing on just using a few techniques that everybody can understand rather than selecting techniques only a few could understand. The importance being that everybody understands how to read and interpret the outputs from these techniques. The Japanese particularly like metric analysis tools that offer easily understood visual representation and you can see most of these satisfy this requirement.

6. Standards

Organisations that separate the work standards from the actual work being undertaken make one of the most profound cultural shifts. If they add to this a measure of performance relating to the work being undertaken then they have immediately established within the culture the prime elements that will support improvements. Standards are in fact the key mechanism that supports quality and efficiency improvement. Change the standards. Run the work through these new standards. Measure performance. Change the standards and re-cycle over and over again. The Standardise – Do - Check - Act (SDCA) continuous improvement (Kaizen) cycle of change.

Standards are necessary to support Quality. So what is quality. There are five steps to quality as applied to products or services.

Step 1 Conformance.

To specification with a commonly used expression being "fit for purpose." Achieving this is focused upon internal organisation processes.

Step 2 Reliability

Does not breakdown and is consistent. Achieving this is focused upon internal organisation processes. The so called "zero defects" strategy extending out to and beyond the customer warranty period.

Step 3 Performance

Speed, capabilities and features. The focus moves from internal to meeting customer expectations. It becomes a features to value equation.

Step 4 Customisation

The variations necessary to meet specific needs. The customer fit.

Step 5 Customer Service

Adding value to the customer relationship

In competitive marketplaces these allow you to gain market share. In non-competitive (eg government) organisations the provision of more for less cost.

In more sophisticated market places (eg USA) market penetration is dependent on Customisation and Customer Service. Their ability to get the conformance and reliability right takes them out of a customer fixing cycle into a customer service cycle. In the car wars between Japan and the USA it was in these areas that the Japanese were particularly effective rather than just the visual appearance and cost to manufacture. They sold on quality of product and service.

In studying quality in a holistic way and working through the five steps to quality you can see a move from internal process improvement to external customer focus. So the early steps focus upon standards, performance measures and problem analysis and rectification. This can be well defined, documented and replicated. The latter steps relate to customer perception and behaviour which moves into intuitive and less predictable areas that prove more difficult to quantify.

Japanese Terminology

Organisations looking to undertake a Japanese approach to organisation change sometimes look to use the Japanese terminology which can bring its own problems because they are terms in the Japanese language. You then not only end up having to learn the change techniques but you also have to master learning and pronouncing the Japanese terms. Organisations have to decide whether to use these

Japanese terms or to use the equivalent Western terminology. Using the Japanese terminology creates clear focus on a new approach but since much of this approach is based upon the original work of Edwards Deming there is a parallel American terminology.

Here we have listed the Top Twenty Japanese Terms. There are many more terms than listed here but these rank as the most commonly used listed in alphabetical order.

01   Andon
02   Five S's (5S)
03   Five Why's
04   Gemba
05   Genbutsu
06   Hoshin Kanri
07   Heijunka
08   Jishu Kanri
09   Jidoka
10   Kanban
11   Kaikaku
12   Kaizen
13   Muda (Severn Wastes)
14   Nagara Cells
15   Newawashi
16   Poke yoka
17   Quality Circle
18   Shojinka
19   Takt Time
20   Yokoten

Andon

Andon
Western Terminology
Warning display
Meaning
A clear visual display that can be observed clearly showing if there are production problems
Process Change
Andon's often have displayed "traffic lights" to indicate if the cell is working or if it has a problem.

Andon is a system to notify management, maintenance, and other workers of a quality or process problem. Often it is a signboard incorporating signal lights to indicate which workstation has the problem. The alert can be activated manually by a worker using a pull cord or button, or may be activated automatically by the production equipment itself. The system may include a means to stop production so the issue can be corrected. Some modern alert systems incorporate audio alarms, text, or other displays.

Five S's (5S)

Five S's
Western Terminology
Standard Operating Procedures (SOP's) or Work Instructions.(WI's)
Meaning
Defines precisely the standards and methods for workplace activities
Process Change
Efficiency by defining best practice

5S is the name of a workplace organisation method that uses a list of five Japanese words: Seiri, Seiton, Seiso, Seiketsu, and Shitsuke. They all begin with the letter "S". The list describes how to organise a work space for efficiency and effectiveness by identifying and storing the items used, and maintaining the area to set standards. It is about setting standards of working practice and communicating these to all employees.

Five S's - Seiri
Western Terminology
Sort, Clear Workplace, Clear Desk Clear Out.
Meaning
Remove all materials not needed in the process.
Process Change
Technique of restricting floor space allocation for the cell.

Five S's - Seiton
Western Terminology
Easy accessibility, Straighten, Configure
Meaning
Ensure everything required is easily accessible
Process Change
Ergonomic layout to minimise movement Applies to simple use of systems

Five S's - Seiso
Western Terminology
Cleanliness, Scrub, Clean and Check
Meaning
Cleanliness, Scrub, Clean and Check
Process Change
Tidiness and clean. Applies to record keeping and databases.

Five S's - Seiketsu
Western Terminology
Planned Cleanups, Planned maintenance, Systemise, Conform.
Meaning
Scheduling regular clean ups.
Process Change
Regular throw away days and clean ups.

Five S's - Shitsuke
Western Terminology
Standardise, Custom and Practice
Meaning
Agreed standards for outcomes and the frequency
Process Change
Defining standards. Includes having standards to define the Five S's

Five Why's

Five Why's
Western Terminology
Root cause analysis
Meaning
Drilling down to investigate the cause of a problem
Process Change
Never accept a fault always determine its root cause

A technique where for an identified problem you ask "why" five times in a drill down style of questioning where the questions are carefully phrased to extract out more detail. It is a way of exploring the cause-and-effect relationships underlying problems.

Gemba

Gemba		
Western Terminology		
Sharp end, coalface, front office.		
Meaning		
Process improvement to be led by those actually doing the work. Focus on where the value is added to the workflow.		
Process Change		
Process design, problem analysis, root cause analysis and problem fixing at the sharp end. Quality Meeting at the sharp end not in a manager's office.		

Gemba is the focus on the sharp end of the business where the work is actually undertaken. It is a clear focus on this area being the most critical to the success of the organisation. It looks to remove the manager's office as being the point at which decisions are made bringing this decision making right to the sharp end. This can be anywhere where the actual work is undertaken so it can be just as relevant to a design office or a sales department as the shopfloor. Team Leaders under Gemba are responsible for Quality (Q) , Cost (C) and Delivery (D) making up the QCD. To achieve this the team Leader manages the five M's

5 M's
Manpower (Person Power)
Materials
Machines
Methods
Measurements

You will have appreciated by now that Japanese culture is very much based upon the use of simple mnemonics really emulating the concepts of mantras. Mantras are short pieces of sacred text continually repeated. The Japanese are very keen on this particular style of communicating what amounts to concise checklists.

Team Leader Supervision Matrix

Quality, Cost and Delivery Maintain and Improve				
Manpower	**Material**	**Machines**	**Method**	**Measure**
Training	Kanban	Total Productive Maintenance	Work Standards	Use of metrics
Morale	Fit for purpose	Set Up Time reduction	Standard Operating Procedures, manuals	Criteria to be measured
Communication	Five S's of good house keeping		Health and Safety	Checks and Controls
Quality Circles				Reporting and interpreting

You can see this matrix of responsibility is inclusive of aspects that may have previously been seen as higher level management responsibilities notably methods and measurement.

The Team Leader should be operating the Kaizen principles. One of the key aspects is the need to undertake observation. Standing and looking at the work being undertaken. Observation should be a regular and dedicated task within the gemba approach.

The gemba approach reflects the idea that whatever reports, measures and ideas are transmitted to management they are only an abstraction of what is actually going on at the sharp end. By viewing everything first hand you not only see the metrics real time but you appreciate the context of the setting. In the western world the Time and Motion Engineers originally employed in manufacturing had the same objectives.

Genbutsu

Gembutsu (at Gemba)
Western Terminology
Artifacts (things) used at the by sharp end.
Meaning
Everything used during the work process. From machines and tools through to components.
Process Change
Defining the artifacts (things) used is part of boundary setting the process.

Genbutsu is the term to describe the tangible objects (ie artifacts) found at the gemba including for example machines, products, tools and the environment itself. The quality and productivity of the work being undertaken will be dependent upon these artifacts

Hoshin Kanri

Hoshin Kanri
Western Terminology
Local Ownership
Meaning
To establish ownership as near to where the work is undertaken as possible
Process Change
The focus is upon the processes that are locally undertaken allowing these to meet the local needs.

Hoshin Kanri is an approach where each operational unit has the capability to define its own strategy but this must be aligned with the organisations overall strategy. It defines a planning methodology in which everybody participates in defining the strategy and planning its implementation. This approach is considered to be more dynamic since the unit will respond to local conditions and be able to make fast local adjustments. It can also have inherent dangers in that local variations can fail to comply with the central objectives. This in Western organisation is often termed the central services verse local services conundrum. The decision as to what to centralise and what to localise keeps organisation managements very challenged. Hoshin Kanni is used at the same time to focus upon the elimination of organisation layers since all levels are more engaged with the overall strategy.

The fundamental need is that the whole organisation marches to the same beat but if central policies destroy local effectiveness then the decision making is better devolved locally. But centralising can improve efficiency because people are more focussed on the same tasks and these can be more automated.

The local needs for change sometimes, termed the hoshins, are inputs from the ends of the organisations nervous system. They need to be implemented quickly to survive but then evaluation is needed to determine if they should be implemented at the centre and broadcast out to other local units. In business terms global organisations operating across many countries need this type of model to be effective in all these different markets.

One of the leading approaches to changing an organisation's culture is to start off with process change under a Kaizen initiative. Trying to effect a cultural change without a real example within the business is difficult. Changing a business process gives real substance and in doing so you can establish a cultural change around this approach. Without such a real piece of work just talking cultural change fails to achieve any real objectives. Remember cultural change is about behavioural change which can be brought out in a business process change project. What is important is the way it is coached with the term coached deliberately used rather than trained. Since coaching is the way to achieve behavioural change.

As you would expect from the Japanese culture to making change they like to use a number of very simple tools that are continually repeated. Not surprisingly they depend on some mnemonics.

Plan - Do – Check - Act (PDCA) Cycle

So you plan or set out an objective. The best is to achieve a metric performance target. In fact just deciding a target in metric terms focuses the mind. Define the current target metric and measure it at present. The do is about applying a change. Then the check is to re-measure. If no improvement recommence the plan - do - check cycle. Once an improvement has occurred then change the standards or processes so the cycle can be repeated consistently. The cultural change is about getting everybody in the organisation to apply these principles to every piece of work.

Standardise - Do - Check - Act (SDCA) Cycle

This cycle focuses on the Standardisation aspect. The standardisation must include the fact that the process has been documented in a way that allows for it to be accurately and consistently repeated. It is the progressive changing of the standard that should generate improved outcomes.

Heijunka

Heijunka
Western Terminology
Line balancing. Production levelling.
Meaning
Achieving a smooth flow.
Process Change
In designing manufacturing processes having more machines when the process is slow to support a smooth end to end flow.

Heijunka is a production levelling process that attempts to keep to a minimum ups and downs in the manufacturing process. It can also link right back to controlling customer demand by reviewing advertising and promotion activity to ensure a consistent demand.

Jishu Kanri

Jishu Kanri
Western Terminology
Team Meetings.
Meaning
Regular short team meeting to support active team working.
Process Change
Regular start of day team meetings

Kaizen depends upon short, yes short, daily review meetings termed Jishu Kanri. They are regular team meetings normally lead by the team leader. They have a very fixed cycle normally daily or more frequently if necessary often before work commences. They work to a fixed agenda and they can be over very quickly if there nothing to report. Creating an event every day for staff to attend to plan and learn has a positive effect quite disproportionate to the time taken which is normally about 15 minutes. It creates a sense of involvement and ownership of the work standards whilst generating higher levels of commitment. It will also generate creative problem solving as fixes to identified problems are shared amongst team members. It is a very focused meeting looking at the specific work the group undertakes for the organisation.

1. What are we doing today and how are we going to meet our output level.
2. What has gone wrong or right since our last meeting.

3. Why did things go wrong or right.

In the manufacturing environment this can be reject focused. In a sales environment sales values and lost sales. In a service environment service levels and complaints. In a professional services environment case reviews including successes and failures.

The prime object is to learn and collaborate on this learning. The term jishu kanri defines a team working in this way.

In the West these has been implemented through a Suggestion Scheme approach where suggestions are rewarded with a prize or money. These were never very effective and they were never given a high organisational priority. The Japanese approach is this improvement process is seen as very much inclusive in the work you undertake. Each worker has a responsibility to contribute to work improvement and change processes with this encouraged through team work.

Jidoka

Jidoka
Western Terminology
Quality Inspection
Meaning
Checking every item to set parameters.
Process Change
In the building of production machines the inclusion in the machine of integral checking devices.

Jidoka is the adoption of a form of automation in which machinery automatically inspects each item after producing it ceasing production and notifying humans if a defect is detected.

Kanban

Kanban
Western Terminology
Just in Time , Visual Systems , Demand Driven.
Meaning
Kaban is a "card" or ticket used to control stock replenishment. Only make what you need and do not carry unnecessary stocks.
Process Change
Materials to point of use just as they are used. Looking for a steady flow rather than stop and start. Reduce inventory stocks. Make sub-components adjacent to the main production line just before required for assembly. In the digital world immediate step to step processing.

Kanban is a card or sheet or token or display or container used to authorise production or movement of an item. Importantly all production and movement of parts and material takes place only as required by a downstream operation. The quantity authorised per kanban is kept to a minimum but it is linked to the demand rate and replenishment times. It does not focus so much on the economics of manufacture (cost per unit) so it is not mass production. It is more about creating a flow and eliminating waste. But the system itself is kept simple to use and low in cost so it is more visually based than computer based.

Kaikaku

Kaikaku
Western Terminology
Project Change
Meaning
A time limited specific change project.
Process Change
It is a major change undertaken in a project way with start and end dates.

Kaikaku means a radical change, during a limited time, of a production system. So it is essentially a project. Kaizen, on the other hand, is continuous minor changes of a certain area of a production system. Kaikaku is about introducing new knowledge, new strategies, new approaches, new production techniques or new equipment. Kaikaku can be initiated by external factors, e.g. new technology or market conditions. Kaikaku can also be initiated when management see that ongoing Kaizen work is beginning to stagnate and no longer provides adequate results in relation to the effort.

Kaizen

Kaizen
Western Terminology
Continuous Change.
Meaning
Change is not viewed as a "project" with a start and end. But it is continuous ongoing process.
Process Change
Process Change is continuous. It never ends.

Kaizen is about achieving continuous improvement. Kaizen is holistic covering all facets of an organisation. Kaizen is executed through working in teams. Teams are cross functional and they must keep a continual focus on the needs of customers or clients. Kaizen is not an initiative with a project like start and end. It is an ongoing organisational culture. This culture is dependent on teams being honest, open and trusting particularly when working across organisational functional boundaries. Kaizen is a learning culture with a focus on ongoing self and team evaluation. Kaizen will support both revolutionary and evolutionary change although the evolutionary is the continuous underlying bedrock on which strategically planned revolutionary initiatives can be instigated.

The Ten Key Principles of Kaizen are:-

Ten Key Principles of Kaizen	
01	Focus on customers or clients.
02	Make continuous improvements.
03	Investigate every problem.
04	Promote openness.
05	Work in teams.
06	Teams must be cross functional.
07	Encourage co-operative team working.
08	Encourage self-discipline.
09	Communicate everything to everybody.
10	Empower every employee.

Muda (Seven Wastes)

Muda
Western Terminology
Waste. Lean Production to avoid waste.
Meaning
Elimination of all waste. Waste can be labour hours or materials or utility resources or unnecessary record keeping.
Process Change
In production environments an easy concept but just as applicable to office settings particularly in respect of wasteful or duplicated record keeping on paper or computer.

Muda is about waste reduction.as an effective way to increase profitability. A process consumes resources and waste occurs when more resources are consumed than are necessary to produce the goods or provide the service that the customer actually wants. It focuses on a heighten awareness and gives whole new perspectives on identifying waste and therefore the unexploited opportunities associated with reducing waste.

Muda is sometimes associated with two other terms mura (unevenness) and muri (overburden) which are about getting the rate of production correct and working to a set "beat".

Muda is analysed into Seven Wastes.

An easy way to remember the Seven Wastes is TIMWOOD.

T: Transportation

Each time a product is moved it stands the risk of being damaged, lost or delayed. It is a cost for no added value. Transportation does not make any transformation to the product that the consumer is willing to pay for.

I: Inventory

Inventory, be it in the form of raw materials, work-in-progress (WIP), or finished goods, represents a capital outlay that has not yet produced an income either by the producer or for the consumer. Any of these three items not being actively processed to add value is waste.

M: Motion

In contrast to transportation, which refers to damage to products and transaction costs associated with moving them, motion refers to the damage that the production process inflicts on the entity that creates the product, either over time (wear and tear for equipment and repetitive strain injuries for workers) or during discrete events (accidents that damage equipment and/or injure workers).

W: Wait

Whenever goods are not in transport or being processed, they are waiting. In traditional processes, a large part of an individual product's life is spent waiting to be worked on.

O: Over-processing

Over-processing occurs any time more work is done on a piece than is required by the customer. This also includes using components that are more precise, complex, higher quality or expensive than absolutely required.

O: Over-production

Overproduction occurs when more product is produced than is required at that time by your customers. One common practice that leads to this muda is the production of large batches, as often consumer needs change over the long times large batches require. Overproduction is considered the worst muda because it hides and/or generates all the others. Overproduction leads to excess inventory, which then requires the expenditure of resources on storage space and preservation, activities that do not benefit the customer

D: Defect

Whenever defects occur, extra costs are incurred reworking the part, rescheduling production, etc. This results in labor costs, more time in the "Work-in-progress". Defects in practice can sometimes double the cost of one single product. This should not be passed on to the consumer and should be taken as a loss.

Nagara Cells

Nagara Cells
Western Terminology
Production Cells or Work Cells.
Meaning
Team of people with a defined assembly or functional outcome. Evolved from manufacturing but is just as applicable to office work.
Process Change
People and Resources are clustered together so human contact is encouraged to achieve a common shared ownership. But they are designed to be able to working at different staffing levels.

Nagara Cells form part of the principles of Cellular Manufacturing an approach in which manufacturing work centres (cells) have the total capabilities needed to produce an item or group of similar items. It is the opposite to setting up work centres on the basis of similar equipment or capabilities, in which case items must move from work centre to work centre. Nagara Cells can also directly feed into the main finished goods production line.

Nemawashi

Nemawashi
Western Terminology
Consultancy. Groundwork. Lobbying.
Meaning
To seek out from all interested parties their views and ideas in respect of a proposed project or idea before it is formally announced..
Process Change
Preparatory work that takes place before the formal announcement. Undertaken effectively it allows for the smooth acceptance by all parties of the project or change when announced.

Nemawashi is an informal process of laying the foundation for some proposed project or change by talking to the people concerned beforehand. It is about gathering support and feedback prior to the announcement. A successful nemawashi process enables changes to be carried out with the consent of all sides. Nemawashi literally translates as "going around the roots" originally meaning digging around the roots of a tree in preparation for its transplant. In Japan, like in most Western organisations, senior management expect to be advised on new proposals prior to any official announcement. If they find out about something for the first time during the meeting then this can cause major problems. It is important to approach these people individually before the meeting to ascertain their reaction to the proposed project or change. This very often results in the original proposal being modified to gain it higher acceptance by all parties.

Poke yoka

Poke yoka
Western Terminology
Fault free design.
Meaning
Design to ensure a component can fit only one way so that it cannot be installed incorrectly. In computer system design this is about data validation and completeness. Stop further data entry until all required data fields entered.
Process Change
Applies to the design stage of products or computer systems. The design is focussed upon preventing faults in the creation process.

Poke-yoka is a Japanese expression meaning "common or simple, mistake proof". It is a method of preventing errors by putting limits on how an operation can be performed in order to force the correct completion of the operation. This is an extremely important concept that is about ensuring that things cannot be assembled incorrectly. It applies to computer records and data as much as physical components. But the principle has to be applied at the design stage.

Quality Circle

Quality Circle
Western Terminology
Fault Finding.
Meaning
Extends out to considering customer and clients inputs.
Process Change
Quality improvement.

Whereas Jishu Kanri is focused upon a specific team and the work it undertakes the Quality Circle broadens things out across teams. Being a member of a Quality Circle is viewed as a promotion and it is often given to someone within the Jishu Kanri meetings that has shown themselves to be very effective. The scope of Quality Circles can extend beyond the organisation to customers or clients. This deals with the feedback to the organisation on the experiences of customers or clients on the product or services offered by the organisation. These are critical inputs to any organisation with considerable effort going into the analysis of the results and rectification activity.

Shojinka

Quality Circle		
Western Terminology		
U Shaped Production Cell Design.		
Meaning		
The design of a production cell so that it can be staffed at varying levels to match the output required. This can be low as one person.		
Process Change		
In the design of the Production Cell it has to be optimised to suit varying staffing levels from one person to the maximum staffing possible thus generating minimum and maximum output levels.		

Shojinka is about the design of Nagara Cells which allows them to run efficiently at different staffing levels with the number of staff increasing as the demand for production grows. But they can be run by one person. Many are designed in a U shape to reduce excessive walking by the operators.

Takt Time

Takt Time
Western Terminology
Production cycle times, Total Actual Cycle Time (TACT)
Meaning
More than just cycle times but balancing to demand.
Process Change
Matching the rate of production to the rate of demand both along a production line cell to cell but at a higher level in line with customer demand.

Takt time is the rate at which product must be turned out to satisfy market demand. It is determined by dividing the available production time by the rate of customer demand. It about setting the beat of the process.

Yokoten

Yokoten
Western Terminology
Knowledge management and sharing
Meaning
The essence is the sharing being everywhere
Process Change
It is not just about sharing but the monitoring of knowledge coming from all sources.

Yokoten means "across everywhere". Knowledge is shared right across an organisation and its suppliers. The sharing of all knowledge across everywhere can only be beneficial. It is a holistic view in respect of knowledge.

Impact of Change Management

The American business that the author worked for produced an automotive component which was sold to a German automotive manufacturer. The contract to manufacture came up for renegotiation. An important part of this process was to review the reliability of the component over the warranty period that the automotive manufacturer offered on their cars. The American business component based upon the German manufactures warranty records showed a failure rate of 150 per million. A Japanese business looking to win the new contract was able to evidence a failure rate of 0 per million, in fact below Six Sigma, for a similar component supplied by them on another contract. It goes without saying that the American business lost the contract. In the automotive industry the adoption of Change Management has now allowed particularly Far Eastern manufacturers to extend car warranties out to 7 years. But on the innovation aspect Toyota instigated a major design change to hybrid electric vehicles only to be out manoeuvred by Tesla who moved to pure electrical vehicles. This re-emphasises the evolutionary Japanese approach against the American revolutionary approach to business. Whilst success comes from combining both the approaches of innovation and quality simultaneously.

The Gurus (Experts)

Organisation Change Management developed after the Second World War led by Edward Deming in the reconstruction of the Japanese manufacturing industries. Edward Deming was called a guru, a term not now in common use, but from the 1950's through to the 1990's this was a common term to describe an "influential teacher and revered mentor." Gurus authored books, ran seminars and setup major consulting businesses to communicate and implement their ideas. Many became very influential and wealthy. They provided a catalyst for change that has now been superseded by the use of the internet technologies.

Gurus supported four distinct waves of organisational change in the manufacturing industries. For each of these four waves the author has selected, based on his opinion, the most prominent gurus contributing to each wave.

The First Wave was in the 1950's was directed by Edward Deming where he was closely associated with two other specialist gurus.

The first guru working closely with Deming was Walter Shewhart (1891-1967) who was a statistician who wrote the "Economic Control of Manufacturing Product" in 1931 introducing the importance of Control

Charts whilst working with Bell Telephone Laboratories. This applied statistics and mathematics to measuring the quality of manufactured parts. It was the foundation of Statistical Quality Control (SQC) which later was extended to be called Total Quality Control (TQC).

The second guru closely associated with Deming was Joseph.M.Juran (1904-2008) who focussed upon the importance of top management being involved in quality improvements and leading any Organisation Change. So Demings, Shewhart and Juran all worked in Japan leading this First Wave through the 1950's. Their consultancy work was very successful in defining the way Japanese Manufacturing was developed which resulted in the global success of Japanese industries.

But possibly more significantly it started the Second Wave of Change Management which was commenced by the Japanese gurus who applied their attention to detail and practical thinking to the concepts established by the First Wave. Like Deming was the lead guru in the First Wave then Taiichi Ohno (1912 -1990) needs to be credited with the lead role in the Second Wave.

Taiichi Ohno was the creator of the Toyota Production System (TPS) which established the principles of Just in Time Manufacturing that extended out in scope to be called Lean Manufacturing. He moved to the Toyota Motor Company in 1943 where he worked as a shop floor supervisor in the engine manufacturing plant rising through the ranks to become an executive. This career path is of significance since he developed a very simple set of shop floor principles applied to the manufacturing process rather than the more complicated academic approach adopted in the West.

The processes were very simple and significantly very visual but possibly more importantly the focus was on the front line workers and in particular their Foremen being the key human contributors to achieving manufacturing success. The so called "Bottom Up" approach rather than the "Top Down" approach so common in the West.

Kaoru Ishikawa (1915 – 1989) expanded upon the management concepts of both Edward Deming and Joseph Juran with an obsessional approach to quality unique to the Japanese culture. He is credited with establishing the importance of Quality Circles and Cause and Effect Diagrams. Quality Circles also developed out of the "Bottom Up" approach that formed a significant part of Taiichi Ohno's Toyota Production System.

The next guru is Masaaki Imai (born 1930) who established the principles of Kaizen. His work as a guru was much later than those listed above and it was more focussed upon the application of the principles to the American marketplace through his Kaizen Institute Consulting Group (KICG). Kaizen is the principle of continuous improvement with a very incremental approach to making changes. Kaizen is also very significant within the Toyota Production System (TPS). So in the Second Wave, initiated by Taiichi Ohno, both Ishikawa and Imai reinforced the importance of vital aspects of the Toyota Production System (TPS).

The Third Wave of gurus takes us back to America where after the outstanding successes of the Japanese manufacturing industries Edwards Deming was requested to bring his ideas and concepts to American industry. These were interpreted into a different type of American approach using the American language because the Japanese terms were unfamiliar and difficult to use. Although fundamentally the underlying principles and concepts were the same it was just the use of different English words to describe them.

Philip Crosby (1926 – 2001) initiated the Zero Defects program at Martin Company on a missile program. In 1979 Crosby started a consulting company called

Philip Crosby Associates. Crosby published a book called "Quality is Free" which was an American attempt to counter the impact on the marketplace of the superior quality of Japanese products.

But possibly the most significant American initiative, which was well publicised, was called Six Sigma. It was introduced by American engineer Bill Smith (1929 – 1993) at General Electric in 1995. A Six Sigma Institute was founded to communicate the techniques of Six Sigma to businesses that joined the institute. A Six Sigma Academy was also established to support a major GE programme to introduce Six Sigma lead by the CEO Jack Welch.

The next guru in the Third Wave is Richard Schonberger who wrote "Japanese Manufacturing Techniques" and "World Class Manufacturing" along with many other books whilst creating a major consulting firm called Schonberger and Associates. He focussed upon extending Change Management out to all parts of a business not just whose relating to the shop floor manufacturing aspects.

The Fourth Wave really was built up on everything that developed in the earlier waves but with the focus very much upon Lean Management. James P Womack became the lead guru with the establishment of the Lean Enterprise Institute. He examined how

Toyota had used their Toyota Production System (TPS) to completely change the world of car manufacture. But he looked to apply this type of Lean Thinking across all manufacturing industries. His book "The Machine that Changed the World" with its memorable title was a best seller which become part of the popular new "Smart Thinking" genre taking over bookshop space.

After these initial Four Waves of gurus the subject of Organisation Change became the responsibility of many varied consultancy businesses many focussed on Lean Management. At the same time Computer Application providers in the late 1980's and early 1990's, particularly in America, started to computerise manufacturing systems. These systems are called Material Requirements Planning (MRP) and Manufacturing Resource Planning (MRP II) systems and there was a whole new set of gurus behind these systems with this subject going beyond the scope of this guide.

Bibliography

The bibliography forms a very important part of this Concise Guide. The Concise Guide was written to give you a very simple introduction to Organisation Change – Japan. To many readers that maybe sufficient detail for your purposes. But other readers may have appreciated the section on the Gurus and now want to go deeper into the subject using the Bibliography.

So this is a visual bibliography detailing the books the author recommends to provide a deeper understanding of the subject. They are on the author's bookshelf. There are hundreds more books on the subject but based upon the author's experience these listed are some of the best available.

These books are all now out of print. But using the internet copies of these books can be easily purchased second hand from the second hand book dealers. Also many of the digital book repositories like "books.google.com" hold copies that can be viewed on line or they can facilitate the purchase of copies.

Although in some aspects the contents of these books may read as a bit dated do not under estimate their value. Because these books were written by the gurus or authors working closely with the gurus the content is a very personal representation of the thought processes these gurus had on the subjects covered.

You get the chance to be engaged directly with their knowledge and understanding of the subjects covered and often more importantly their rational and reasoning behind the solutions they recommended.

It has to be said with all the technological advances through smartphones, tablets and personal computers using the internet with masses of information and knowledge readily available through a variety of search and multi-media techniques nothing has been found to be effective as a book to communicate knowledge. With the best being a paper based book you can hold in your hands without any technology interfering with your thought processes. Pure knowledge transfer from book to brain.

Note on ISBN Numbering

Most of these books use the older original ISBN-10 Standard Reference Code but some re-prints may use the newer ISBN-13 Standard Reference Code.

Out of Crisis. Edwards Deming. 1982.
Cambridge University Press. ISBN 0521305535

This is classic book on the subject written by the leading guru Edwards Deming. If you want to get to the core of this subject this is the book to read. It reads in a "dated" way in some areas but it does have the original ideas on how Edwards Deming's mind thought through this subject. The broad range of his subjects is enlightening with surprisingly his focus more on people than on statistics.

The New Economics. Edwards Deming. 1993
Massachusetts Institute of Technology.
ISBN 0911379053

A simpler less complicated book than Out of Crisis. Double spaced and larger text font makes for easy reading. Good introduction to systemic thinking in Chapter 3. Excellent Chapter 8 on "Shewhart and Control Charts". This book is a more up to date look at change with a more computer focussed approach although it is too old to reflect on the impact of internet. It is still in 2019 a thought provoking book to read since many things do not change over time.

Keys to Excellence. Nancy R. Mann. 1989
Mercury Business Books. ISBN 1852510978

This is highly recommended. It benefits from another author, Nancy R. Mann, writing about the Edwards Deming philosophy. This gives a more simplified perspective on his Deming Philosophy. It uses many real world examples to illustrate the points being made by Deming. The Chapter 1 on "Why it happened in Japan and not in the US." framed much of my thinking in the writing of this book. The culture in Japan values perfection and quality in goods and services beyond that of Western cultures.

Juran on Leadership for Quality. J.M.Juran 1989
The Free Press. ISBN 0029166829

Chapter 7 on Operational Quality Management is ahead of its time with a focus upon the importance of processes and process flowcharts. But one of Juran's key areas is getting an organisation's culture changed to focus on quality with him reinforcing the fact that this has to come right from the top of the organisation. The problem is often getting the top level in the business to get quality motivated and show quality leadership. The essence to his thinking is quality is about people.

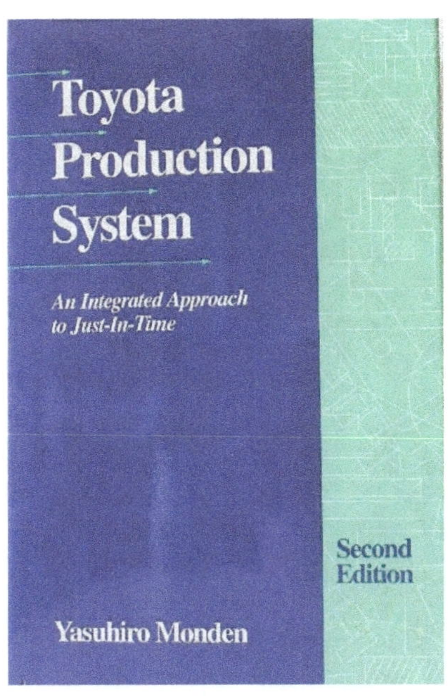

Toyota Production System. Yasuhiro Monden. 1993
Institute of Industrial Engineers. ISBN 0898061296

The classic case of an author, Yasuhiro Monden, doing an excellent job of explaining Taiicho Ohno's Toyota Production System in a Western style of writing. Heavily focussed upon the detail of the shop floor manufacturing activities. Ideally suited to the reader that works in the manufacturing industry at the shop floor level. Includes loads of detail and diagrams looking into every small detail of manufacturing. Even if you are not into manufacturing the detail is worth reading to illustrate how thorough is the Japanese approach.

What is Total Quality Control? Kaoru Ishikawa 1985
Prentice Hall. ISBN 0139524339

All credit to the translator David j. Lu who has in the process of the translation of Kaoru Ishikawa original Japanese book achieved the simplified wording that makes it very readable. It is focussed upon Total Quality Control (TQC) considering much broader aspects of quality across a business than just the shop floor aspects. It is a very readable style without a lot of the mathematical aspects so common in some books on TQC. Easy to read it holds your attention well and you can acquire a lot of knowledge with minimum effort.

Kaizen. Masaaki Imai 1986
McGraw- Hill. ISBN 007554332X

An excellent detailed book going into depth on the subject of Kaizen. A very diagrammatic book with many documented examples illustrated and drawn from a real manufacturing settings. Ideal for the reader that wants to look at real world examples covering the application of Kaizen. It focusses upon the single measure on how successful Kaizen has been in a business as being Customer Satisfaction. It emphasises that no matter what management does it is to no avail if it does not lead to increased customer satisfaction. So management should try ringing their own call centre's ?

Six Sigma. Mikel Harry and Richard Schroeder 2000
Doubleday   ISBN 0385494378

If you read only one book on Six Sigma make it this one. It reads so well and it has an excellent structure. It combines theory with excellent examples. Not surprisingly General Electric (GE) gets a lot of coverage but other companies are well represented. My copy is covered with highlighted sentences so it shows how often the content triggered my thought processes. The key marketing element of this approach was to use the name Six Sigma which is a failure rate of 3 per million.

World Class Manufacturing Richard Schonberger 1986
The Free Press   ISBN 0029292700

Author of "Japanese Manufacturing Techniques" whilst in this book he looks at nearly 100 American Corporations and how they have applied Just in Time (JIT) Production and Total Quality Control (TQC). The book has many diagrammatic examples and goes into considerable detail. You may look at the date of publication as being 1986 but many manufacturing businesses in 2020 have still not applied these ideas so reading it 34 years later it still a valuable exercise.

Building a Chain of Customers R.J. Schonberger 1990
Hutchinson   ISBN 0091745993

The book looks at the transformation of the whole organisation so the focus ceases to be department centric but about the whole organisation serving the customer. It follows the idea that each department in an organisation acts as both a supplier and customer to other departments with the overall effectiveness of this effecting how the real customer is served by the whole organisation. Although dated in terms of content it includes many business examples where the improvement messages are just as valid in 2020.

Lean Thinking   Womack Updated 2003
Free Press   ISBN 0743231643

A classic book on Lean Business Transformation applicable to any business not just manufacturing. Includes many of the concepts of Lean Management going back to the fundamentals but looks at how they have been applied in many different businesses. Rightly acknowledges that Toyota leads the global car industry with many other businesses looking to catch up with their advanced lean thinking.

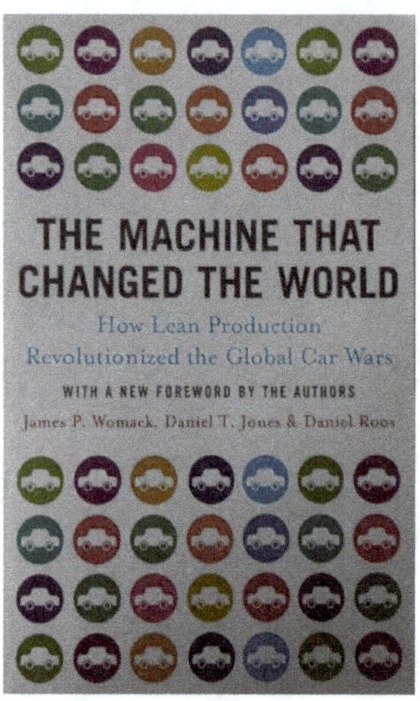

The Machine that Changed the World. J.P. Womack 2007
Simon & Schuster   ISBN 1847370551

Although this is the last listed in this Bibliography it was originally written in 1990 and it really serves to summarise how the work of Deming and all the other gurus completely changed manufacturing industries. If you do not read any of the others listed then you must read this one. It makes for very easy reading and it should motivate you to want to read many of the other books listed. Once again the success of Toyota and their approach to Lean Management is acknowledged as the approach that changed the world of manufacturing.

Endnote

This book has provided you with a brief introduction on how Organisation Change is undertaken within Japan. It has introduced you to some of the Japanese cultural aspects that have made them so successful particularly in their manufacturing industries.

More than anything else, if I have been an effective author, this should have now encouraged you to want to read more on this subject. Emulating the latest trends in information retrieval and knowledge acquisition this book has been written to give you a fast introduction to Organisational Development – Japan. You should now use Google and Wikipedia to lookup the key words within the text allowing you to read up more comprehensive detail. Or use Amazon Books to obtain further copies of books on the subjects covered in this book.

This book is also once again an acknowledgment of Edwards Deming's excellent work and reminds us of the fact that the Japanese took on board his ideas 30 years earlier than the American's. It worth noting that the major changes in the world are initiated by the thinkers but they need listeners to act on their ideas. It goes without saying that the Japanese are very good listeners.

But it has to be noted that although the Organisation Changes outlined in this book align themselves very well with the Manufacturing Industries, producing real physical products, not all of techniques are as applicable to the so called new Digital Industries.

The Digital Industries, representing the so called Fourth Industrial Revolution, require a much more unstructured approach to the creation of new digital solutions. It is their creation that is critical because this is where new ideas are invented and explored resulting in new digital solutions. But once the digital solutions have been created their reproduction and distribution lacks the complexity of the conventional Manufacturing Industries.

But throughout this book the focus placed on both Teamwork and Quality is just as applicable to both the Manufacturing and Digital Industries.